DATE DUE

Fecha Para etorn

D0123941

W9-AOP-871

Hide and Seek

Hide

and Seek

Creatures in Camouflage

Phyllis J. Perry

A First Book

Franklin Watts
A Division of Grolier Publishing
New York   London   Hong Kong   Sydney
Danbury, Connecticut

Photographs ©: Animals Animals: 19 (Derek Broomhall/OSF)), 28 (C. Dani/I. Jeske), 8 (Michael Fogden), cover, 31 (Zig Leszczynski), 40 (Ralph A. Reinhold), 11 (Anup Shah); Norbert Wu Photography: 48, 52; Photo Researchers: cover (Alan L. Detrick), 2, 12, 22, 26, 37 (Gregory G. Dimijian), 43 (Jerry L. Ferrara), 14 (Michael Giannechini), 46 (J.A. Hancock), 15 (David Hosking), 29 (A.B. Joyce), 44 (Tom & Pat Leeson), 32 (Jeff Lepore), 51 (Fred McConnaughey), 38 (Anthony Merceca), 17, 24 (J.H. Robinson), 33 (John Serrao), 42 (Jim Steinberg), 55 (Andrew G. Wood); Wildlife Collection: 35 (John Giustina), 50 (Chris Huss), 21 (Charles Melton).

Library of Congress Cataloging-in-Publication Data

Perry, Phyllis Jean
    Hide and seek: creatures in camouflage/ Phyllis J. Perry
    p. cm. — (A First book)
    Includes bibliographical references and index.
    Summary: Discusses the importance of camouflage in the animal kingdom, describing different types of disguise including disruptive coloration, countershading, mimicry, and masking.
    ISBN 0-531-203-9
    1. Camouflage (Biology)—Juvenile literature.   [1. Camouflage (Biology) 2. Animal defenses.]I. Title   II. Series.
QL767.P47   1997
591.47'2—dc21   96-37290
                                                      CIP
                                                      AC

For Melissa Stewart
who nurtures ideas until they spring to life.

Special thanks to
Kathy Carlstead, Ph.D.,
National Zoological Park, Washington, D.C.,
for her helpful comments and suggestions

Contents

When a mantid sits quietly on a plant, such as this colorful bougainville, it is almost invisible.

Chapter 1

Types of Camouflage

A sharp-eyed bird flies low over a pink flowering bush in a Malaysian jungle. It is looking for a tasty insect to eat. Seeing nothing, the bird continues on its way. A few minutes later, a graceful butterfly dips down over the flowering bush. It lands on what looks like a lovely pink blossom. Almost instantly it is seized and eaten by an orchid *mantid*.

The pale pink orchid mantid was there all the time. But it was invisible to both the bird and the butterfly because of its *camouflage*. The mantid's entire body is exactly the same shade of pink as the flower it hides in. Even its eyes and antennae are pink.

The orchid mantid is just one of more than 1,500 species of mantids. The front legs of all mantids are used to seize *prey* in a viselike grip, hold the victim firmly, and tear it apart. Many mantids blend in with the vegetation on which they live. Their coloring helps them hide from their enemies and lie in wait for their victims.

For the most part, camouflage is a trick that involves the onlooker's eyes and brain. If an animal blends into the background and remains perfectly still, it may not be noticed by an observer. In other words, the animal will be seen by the eyes of the observer, but it will not be recognized by the observer's brain.

There are many different types of camouflage. In most cases, the color and pattern of an animal's fur, skin, or feathers makes the difference between eating and being eaten. As you will see, some animals can even change their body coloring from one season to the next or when they feel threatened by an enemy.

Colorful Camouflage

Tigers are just one of the thousands of animals that rely on their coloring for camouflage. The tiger, the largest member of the cat family, may weigh more than 500 pounds (225 kg). It is a *carnivore* and will eat almost any animal it encounters. Although tigers prefer deer, antelope, wild oxen, and wild pigs, they will also eat peafowl, fish, scorpions, monkeys, tortoises, and frogs.

The *disruptive coloration* of the tiger's vertical stripes helps to break up its outline. As a result, a tiger is almost invisible as it creeps through tall grasses or forest shadows.

A tiger's stripes make it difficult to spot as it creeps through tall grass.

It is late afternoon on a grassy plain in southeast Asia. As a hungry female tiger approaches a water hole, it spots a large group of brightly colored peacocks taking a drink The mighty tiger slowly moves closer and closer, but the birds remain completely unaware of its presence. Suddenly the tiger bounds toward its target, grasps one of the birds, and pushes it to the ground.

The tiger was so intent on the peafowl that it failed to notice a helpless young muntjac—a type of deer—hiding nearby. The muntjac would have made a better meal. A muntjac's reddish-brown fur is dappled with white spots. When these deer stand still, they blend in with the tall grass.

Because peafowl are brightly colored, they are an easy target for tigers and other *predators*. Most other birds are much more difficult to spot. Patches of gray, brown, and white break up their outlines, making them harder to see. Some birds also have an eyestripe, which helps conceal their eyes.

Many *reptiles* also take advantage of disruptive coloration to hide from their enemies and capture their prey. One example is the Gaboon viper, which lives in the rain forests of central Africa. It may be up to 6 feet (1.8 m) long and weigh as much as 18 pounds (8 kg). The viper's long fangs, potent venom, and camouflage make it very dangerous. The scales on its sides and back form patterns that blend well with the leaves on the forest floor.

Stripes and spots are not the only coloring tricks used in nature. Some animals can change the color of their entire bodies to match their environment. If falling snow changes the color of their *habitat* from brown to white, these animals do not panic. They know exactly what to do. They shed their brown summer jacket, and replace it with a heavy white winter coat.

As winter approaches, the fur of the snowshoe hare changes from grayish-brown to white, so that the hare will blend in with the snowy landscape. When spring returns, the hare begins to shed its winter coat. By summer, the hare wears a fine, grayish-

The pattern on the scales of this gaboon viper help it blend in with the leaves littering the ground.

It takes 10 weeks for the snowshoe hare's brown summer coat to turn into the white winter coat seen in this photograph.

brown coat. The snowshoe hare's ears and feet remain white all year.

Concealing coloration is the term that scientists use to describe camouflage that helps an animal match its environment. In the case of the snowshoe hare, this involves changing color with the seasons. Many animals do not have to change their body color to resemble their surroundings. The pink color of the orchid mantid is a perfect example.

Sunlight and Shadow

Some animals rely on contrasting colors to help them hide. Their bodies are dark on top and light underneath. Sunlight shining down on the darker color tends to lighten it, while shadows on the underside of the animal's body make it look darker. As a result, the two colors blend. This effect, called *countershading*, is particularly useful for animals that live in water.

Although the bontebok is about the same size as a deer, countershading makes this African antelope look smaller. The top of its coat is a rich brown color with a white patch on the rump. Its face, legs, and belly are white. As the hot sun beats down on the African savanna, the brown appears lighter. Meawhile, the shadow cast on the animal's underside makes it look darker. As a result, the bontebok blends in with the grass it is grazing on.

There is another way that animals can use shadows to trick predators. Hunting animals often spot a victim because they notice its shadow. If a prey animal can make its shadow disappear, it may be able to save its own life. The closer an animal stays to the ground, the smaller its shadow will be. For example, when a racing crab senses danger, it races to a hollow and lies flat on the sand to hide its telltale shadow.

Countershading helps this bontebok hide from predators. Notice that its back is dark brown and its belly is white.

Other Tricks of the Trade

Some animals have yet another technique for blending into the background. They pretend to be something else. A harmless flower fly has the same size, shape, and coloring as a nasty, stinging yellow jacket. Because the two insects look so similar, their enemies often mistake flower flies for yellow jackets. Because the predators do not want to be stung, they avoid both yellow jackets and flower flies. This type of camouflage is called *mimicry*.

Weevils use a technique called *masking* to protect themselves. They hide themselves by attaching small plants, animals, or other items to their bodies. Some types of weevils mask themselves by growing little gardens on their back. The dirt and tiny plants growing on top of them make them very difficult to spot.

Masking and mimicry, countershading and coloring help hide prey animals from their enemies, but they also allow the predators to catch their victims by surprise. This is true of spiders and insects, reptiles and *amphibians*, *mammals*, birds, and ocean creatures. In the wild, hide-and-seek is no children's game. It is a matter of survival.

By mimicking the colors and stripes of a wasp, a harmless flower fly may trick its enemies.

Chapter 2

Spiders and Insects in Camouflage

The crab spider doesn't build a web to catch its food. It doesn't have to. It can change its color to match the petals of whatever blossom it decides to hide in. As a result, it is very difficult to see a crab spider as it waits for dinner to arrive.

In the summer, the crab spider is usually white. However, it turns light pink when it sits inside a pink rose. Soon after climbing into a yellow daisy, the crab spider becomes pale gold. This spider is taking advantage of its concealing coloration.

Because its victims do not notice the spider, they often get so close that the crab spider can grab them. Before a helpless fly knows what has happened, the spider has used its sharp fangs to inject deadly venom into the fly.

The female sand wolf has dark rings on its legs and a pale body sprinkled with black and brown marks. When one of these

This crab spider has altered its body color to match its surroundings, a pink dogrose. If a fly comes too close, it's in for a nasty surprise.

spiders stops moving, it is almost invisible among coastal sand dunes. Predators, such as the pompeliid wasp, often have trouble spotting this spider.

A sand wolf lives in a silk-lined burrow. If the weather is poor, or the spider senses danger, it can spin a silk curtain over the opening of its home in just a few seconds.

How Insects Hide

Many insects make use of disruptive coloration. The small ambush bug commonly found in the eastern United States is one example. Its greenish-yellow color and a pattern of black bands, along with its spiny armor, conceal it well. Nearly invisible, it sits on a goldenrod waiting for nectar-seeking insects to come within its reach. The ambush bug uses its powerful, toothed forelegs to kill small prey—and even insects considerably larger than itself.

The wings of the peppered moth have a black-and-white speckled pattern that resembles local tree trunks. There are both light-colored and dark-colored forms of this moth. In England, the light-colored form was once more common. But as industries grew and trees near factories were darkened by soot and smoke, larger numbers of darker moths were found. In some areas, they have almost completely replaced the light-colored form.

Many insects use *disguise* and mimicry to hide from hungry predators. For example, stick insects resemble small twigs. They have tiny heads, and long, thin legs and antennae. Most are light brown, gray, green, or a combination of these colors. The common walking stick, which is found throughout the eastern United States, looks like a pale-green and brown twig. It is almost indistinguishable from leaves and sticks scattered on the ground.

This ambush bug's colors are so close to the colors of the goldenrod that it is well hidden from its predators and its prey.

No, you are not seeing a little alligator in a tree. This strange insect, often called an alligator bug, is only about the size of a peanut.

Brazil is the home of a strange insect called the lantern fly, but since it isn't a fly and doesn't give off light, its popular name—the alligator bug—fits it much better. This insect's light-weight head, which is about the size of a peanut, makes up about one-third of its total body length. It looks like a tiny alligator head complete with eyes, nostrils, and teeth. The insect's real eyes lie behind this "false head." The alligator bug is a peaceful plant eater, but its fierce looks discourage predators from attacking it.

At rest, the kallima or Indian dead-leaf butterfly, displays the light-brown undersides of its wings, which resemble a dead leaf. These undersides have dark lines that look like a leaf's midrib and veins as well as a spots that look like the discolorations found on many decaying leaves. When this butterfly spreads its wings and flies, it reveals the top sides, which have a spectacular blue and orange pattern.

The name of a large moth family, Geometridae, means "earth measurer." The caterpillar stage of this moth's lifecycle is often called an inchworm or measuring worm. According to a popular children's story, if you sit still and let an inchworm "measure" you by inching its way across your body, you will soon receive a new outfit.

Many types of inchworms have slender bodies and irregular shapes that resemble twigs or the stems of leaves. These inchworms often grasp a branch with just their hindmost legs and stretch their bodies out at an angle from the stem. This

makes them look like just another twig sticking out from the branch.

Some insects are not bothered by predators, not because they are difficult to spot, but because they resemble an insect that stings or tastes bitter. For example, one type of cockroach has orange and black spots that make it look like a ladybug. Since ladybugs do not taste good to many insect-eating creatures, this mimicry may save the cockroach's life.

Spiders and insects are not the only creatures that use coloring and trickery to fool their predators and their prey. In the next chapter, you will see how animals as different as snakes and salamanders use camouflage to protect themselves from their enemies and sneak up on their victims.

This caterpillar, which is the larva of a large moth, disguises itself as a twig by holding onto a branch with its hindmost legs.

other eye swivels until it finds the victim. Once the chameleon has both eyes on its target, it can accurately judge the exact position of its prey.

Scientists do not think that chameleons deliberately alter their color for concealment. Instead, the color change is a reaction to changes in light, temperature, and emotion. In fact, most chameleons don't need to change color to hide from their enemies. Their disruptive coloration and compressed body shape make them difficult to notice among shadows and leaves. In other words, their usual coloring already matches their typical surroundings.

The jewel chameleon of Madagascar is one of the most strikingly colored animals on earth.

A frilled lizard of Australia hopes to frighten away predators by opening its mouth in a threatening way while raising the fold of skin around its neck into a colorful ruffle.

The jewel chameleon of Madagascar is brightly colored. It has white, black, blue, and red stripes and spots. But most of its body is a leafy green, and blends in with foliage. When this chameleon sits very still, it is well hidden from its enemies.

Most of the time, the jewel chameleon waits patiently for an insect to come within reach of its sticky tongue. When it spots a potential meal, the chameleon shoots out its tongue, snatches the insect, and quickly pulls the victim into its mouth.

While concealing coloration works well for chameleons, other lizards have a very different way of protecting themselves. They make themselves as visible as possible and try to discourage approaching enemies. The Australian frilled lizard tries to frighten predators with sudden flashes of color.

When this lizard becomes frightened or angry, a collarlike fold of skin around its neck stands up straight and forms an enormous ruffle. At the same time, the lizard opens its mouth, which is bright yellow inside. This shocking display often scares the lizard's predators. If it doesn't, the lizard stands up on its hind legs and runs off. When the lizard no longer feels threatened, its collar relaxes. Most of the time, this collar hangs down around the lizard's shoulders like a cape.

The leaf-tailed gecko of Madagascar is about 12 inches (30 cm) long. Jagged frills on its body help disguise its outline. The tail, limbs, and sides of this gecko are edged with thin flaps, which reduce this animal's shadow. The flaps also help the gecko blend in with its background.

As the leaf-tailed gecko moves among the *lichen* patches that grow on the bark of trees, its body color changes to match its lichen backdrop. After just 20 minutes, it is nearly impossible to distinguish this gecko from its surroundings.

Lizards are not the only reptiles that take advantage of camouflage. In fact, the Saharan sand viper uses its own colorful tricks to catch lizards—its favorite meal. During the day, sand vipers bury themselves in the sand. When a lizard approaches, the sand viper raises the tip of its black-and-white banded tail above the sand.

The lizard may think that the black-and-white wiggling spot is some type of tasty insect larvae, and move toward it. When the lizard is close enough, the viper strikes.

An Egyptian sand viper is well camouflaged as it waits to attack unsuspecting prey.

Some harmless snakes are protected from their predators because they mimic a venomous one. Southeast Asia is the home of a deadly coral snake. Its brightly colored rings and red belly warn other animals that it is dangerous. A harmless snake from the Colubridae family has similar coloring.

These two snakes not only look alike, but also behave the same way when disturbed. Each snake raises the tip of its tail to show enemies its "false head," and rolls onto its back to reveal the black-and-white bands on its underside.

Nomadic Newts, Sly Salamanders

Reptiles are not the only creatures that use bright colors to warn their enemies to stay away. The red-spotted newt is an amphibian that usually lives in ponds in the eastern United States. The adults, which are dull green with a row of red spots on each side, lay their eggs in the water.

After a few weeks, gilled larvae hatch from the eggs. In most cases, these larvae grow into green newts, just like their parents. But in some places, newt larvae behave quite differently. They turn a bright orange-red color and leave the pond.

The red eft (above) and the red salamander (facing page) are not identical, but they resemble one another so closely that some predators are fooled. Birds may pass up a salamander thinking that it is a red eft, which has bad-tasting skin.

These red *efts* spend 2 or 3 years on land. When they return to the water, they turn dull green. The bright coloring of red efts warns their enemies to stay away. Any predator that tries to eat an eft will quickly learn that its skin contains a potent poison.

Red salamanders take advantage of the presence of their newt neighbors. Scientists believe that the red salamander is a mimic of the red eft. While the two creatures do not look identical, they are similar enough to fool many birds. The red salamander also imitates some of the red eft's behaviors. Both amphibians lift and wave their tails when they feel threatened.

Throughout the animal world, different species take advantage of the same basic camouflage techniques. Mimicry is the method used by red salamanders and flower flies, stick insects and inchworms. In the next chapter, you will discover how mimicry is used by a squirrel that lives in southeast Asia.

Chapter 4

Mammals in Camouflage

The spots, stripes, and blotches on the fur coats of some mammals help them hide from predators and sneak up on their prey. The spotted leopard lives on the African savanna, where its spots allow it to creep unseen through tall grasses. This leopard is also nearly invisible as it lies on tree branches and waits for prey to pass below. The dark coat of the black leopard matches its surroundings—the dense jungles of Malaysia, Java, and India.

If you look at zebras in a zoo, you will notice that their stripes stand out clearly. Believe it or not, these stripes help zebras blend with the tall grasses of Africa. Their disruptive coloration conceals their body shape, so enemies cannot spot them from a great distance. When a lion sprints after a running herd, the zebras' stripes confuse the lion. The predator has trouble seeing where one zebra ends and another begins. The lion is unable to pick out one target to chase. This is particularly true

The young tapir has very different coloring from its mother. Spots and stripes help many young mammals to blend in with their surroundings.

at dawn and dusk, when lions and other predators do most of their hunting.

Because young mammals are more vulnerable to attack, many have special protective coloration. While adult deer are usually solid brown, fawns have spots that help them hide among forest shadows. Adult tapirs are also solid brown, so they can't be seen as they feed at night. Young tapirs are brown with white stripes. These stripes allow them to blend into bushes on the forest floor.

The fur of many mammals helps them to blend in with their surroundings. From a distance, it is difficult to distinguish mountain sheep from the rocks on a steep mountain cliff. Their dark grayish-brown coats match the rocks that they clamber among as they look for mountain grasses and plants to eat.

Some animals come in a variety of colors. The color of a particular individual depends on its environment. For example, pocket mice may be black, white, or light yellow. Those that live on the black lava rock of New Mexico are almost black. At White Sands National Monument, the pocket mice are almost white. In a nearby sandy desert area, they are light yellow. This type of camouflage is called *polymorphism*.

Mimicry is less common among mammals than it is among other groups of animals. One species of squirrel found in southeast Asia is not bothered by most of its potential enemies because it looks so much like a tree shrew. Both animals have a slender tail and long nose. Flesh-eating animals avoid the shrew because it has an unpleasant taste.

When a sloth hangs upside down among the branches of a tree, it is very difficult to see. That's because its fur matches the tree trunks commonly found throughout the tropical forests of Central and South America. The sloth does not rely solely on concealing coloration, however. It also takes advantage of masking. The algae that grow in its fur during the wet season resemble the lichen that grows on the trees when more water is available.

Juvenile two-toed sloths hang from the trees in the Monteverde Cloud Forest Reserve of Costa Rica.

While masking is used by relatively few animals, disruptive coloration is extremely common. Spots, stripes, and unusual patterns protect leopards and zebras, peppered moths and ambush bugs from their enemies. Disruptive coloration is also very common among birds. Read on to learn how a whip-poor-will's speckled feathers make this bird difficult to spot. Other birds, such as the American bittern and Jack Snipe rely on their stripes for safety.

Chapter 5

Camouflage Among Birds

The whip-poor-will is *nocturnal*—it hunts for insects at night and rests on the ground during the day. Because the whip-poor-will spends most of its time on the ground, it is in constant danger of being eaten by foxes, skunks, weasels, and snakes. Luckily, the feathers of whip-poor-wills and most other ground birds are speckled or have brown and black patches, so they blend with the dirt, rocks, and twigs that cover the ground.

The brown creeper cannot fly very well, so it "creeps" up and down tree branches in search of insects. It's long, slender beak is perfectly designed to probe into small crevices along a tree's bark.

Because it spends so much time on the ground, the whip-poor-will must rely on its speckled coloring for protection.

This female cedar waxwing is feeding wild berries to its young. Notice the mother's dark eyestripe.

When this bird remains very still, it is hard to spot. Feathers with streaks of brown cover the creeper's wings and back. These markings blend with the bark of most North American trees.

But, even when a bird is sitting perfectly still, its round black eyes may stand out. The cedar waxwing is one of the many birds that has eyestripes to camouflage its eyes. A cedar waxwing's dark eyes are difficult to see because the bird has a black stripe that begins just below a tuft of feathers at the top of its head. The stripe sweeps across the waxwing's face and blends with its black beak.

The cedar waxwing's brownish plumage has a silky texture. Its gray wings and white throat patch distinguish it from other birds. Cedar waxwings are found throughout the United States and Canada. Their favorite foods include wild cherries and mulberries.

The American woodcock is also known for its prominent eyestripe, which extends from each side of its beak and passes directly over its eyes. The woodcock also has four rust-colored bars across the top of its head.

This bird, which is about 11 inches (28 cm) long, lives in the wet thickets, moist woods, and bushy swamps of the midwestern United States. It usually feeds in lowlands, where it can bore into the soft mud for worms and other small creatures. The woodcock nests on the ground and lays tan or reddish-brown spotted eggs. Because the woodcock's upper body is a blend of rust, black, brown, and gray, the bird is difficult to spot among fallen leaves.

One of the strangest birds in the world is the Australian tawny frogmouth. Like the whip-poor-will, it hunts at night. The frogmouth catches creeping insects as well as frogs and other small animals with its huge mouth. During the day, the bird sits in a tree with its eyes closed and its beak pointed. This posture along with its mottled-gray coloring makes the frog-mouth look like a broken tree branch.

After a night of hunting, the Australian tawny frogmouth roosts high in trees. Its mottled gray coloring makes it look like a tree branch.

A bittern points its bill at the sky and does its best to look like just another slender reed growing in the swamp.

Bitterns are wading birds that live in wet meadows and marshes as well as reedy swamps. During the day, they search for frogs, small fish, meadow mice, and other small animals. When a bittern locates a victim, it spears the prey with its sharp bill.

The American bittern has stripes on its breast. These stripes help camouflage the bird because they resemble the long, thin plants that grow in the swamps where it hunts, builds its nest of reeds, lays its eggs, and raises its young. When a bittern senses danger, it lifts its head and points its bill toward the sky. As a result, it looks like another reed. The bird even sways back and forth as the reeds blow in the wind.

The Jack Snipe, another marsh bird with stripes, has light-and-dark-brown stripes that run the entire length of its body. When the snipe lands, it positions itself so that its stripes are oriented in the same direction as the reeds around it.

A rock ptarmigan is slightly larger than a pigeon. It is found in the mountains of Alaska, Canada, and in the Rocky Mountain region of the United States. This bird goes through three color changes each year.

In the summer, the ptarmigan's feathers are covered with brown speckles that match the dirt and rocks that cover the ground. In the winter, its feathers turn white, so it can hide in the snow. And in the spring, patches of brown develop among the white feathers, so the bird blends in with the patches of ground that become visible as the snow begins to melt.

While disruptive coloration is the most common type of camouflage among birds, some use their concealing coloration to fool predators. The blue-crowned hanging parrot, is a good example.

This parrot has a yellow breast, green wings, and a blue spot on the crown of its head. When these parrots hang upside down to roost, they look just like a bunch of leaves. Sometimes, they even feed while hanging upside down.

This rock ptarmigan has its summer plumage, which allows it to blend with rocks. In winter, its feathers turn white, and in spring there are patches of brown among the white feathers.

The beautiful green coloring of Livingston's turaco helps hide it from predators in the tropical rain forests of Africa.

Livingston's Turaco, which lives in the tropical rainforest of Africa, is also a very colorful bird. The turaco's bright scarlet and green feathers are very unusual. Although you might think that its brilliant coloring would make the turaco an easy target for hawks and other predators, its green feathers match the color of the vines and ferns in which it lives. Turacos flash their red flight feathers at one another during courtship.

In some cases, male and female birds of the same species have very different coloring. The male wood duck is one of the most colorful birds. Its feathers are rust, gold, white, green, red, and blue. In comparison, the female is a rather plain, splotchy-brown color.

The males are brightly colored to attract females, but unfortunately, their vivid coloring also makes them more visible to their enemies. The dull coloration of the females help them avoid detection by predators when sitting on a nest full of eggs.

By now you know how important it is for animals to blend into their surroundings. Gray and brown splotches make many birds almost invisible on the forest floor. A lion's golden coat matches the dry grasses of the African plain. The spots of the fawn and the leopard make them hard to see. Snakes, insects, and lizards rely on colorful tricks too. Camouflage is important for all land animals. In the next chapter you will see how animals living in the ocean also depend on their coloration to avoid their enemies and sneak up on their prey.

The blue-ringed octopus uses its deadly venom to para-
lyze the respiratory system of its prey. It pulses blue
rings when it is angry.

Chapter 6

Camouflage in the Ocean

A blue-ringed octopus quietly moves into a small hiding place in a coral reef off the shore of Australia. Within minutes, the octopus changes the color of its body from blue-green to a mottled brown to match its new surroundings. Some octopuses can even change the texture of their skin. For example, when these octopuses rest on a rock, their skins get bumpy.

When an unsuspecting crab approaches, the octopus swiftly grabs its victim with one of its long tentacles. Many octopuses have a very unusual way of tricking their predators. When an enemy approaches the octopus releases puffs of an ink-like substance into the water. The inky cloud confuses the predator, giving the octopus a few seconds to escape.

The cuttlefish can also change color to camouflage itself. When the cuttlefish is out hunting, different colors and patterns move across its body. The cuttlefish accomplishes these color

This flesh-colored cuttlefish is difficult to see among a group of rocks near the Solomon Islands.

changes by opening pouches, called *chromatophores*, that contain yellow, orange-red, and black pigments. The colors can be used alone or combined to create a variety of different shades.

Many flatfish, including flounders, rest on the ocean floor. They are white underneath, but their backs change color to match the pebbles or sand on which they are resting. In about 2 hours, not only their color but their patterns change to match their background. Carefully camouflaged, they wait for unsuspecting prey to swim by.

There's more than sand here. Can you make out the shape of the young peacock flounder lying flat on the sandy bottom of the ocean floor?

The batfish cannot alter the color of its body, but it doesn't have to. It takes advantage of a different type of camouflage. This fish, which often swims near submerged mangrove trees, looks just like a yellow mangrove leaf. If attacked, the batfish swims down to the bottom and lies on its side among the leaves. Its disguise is so effective that predators are often unable to find the batfish. When the coast is clear, the batfish swims up to the surface again.

Some fish, like the copperband butterfly fish, have an eye on each side of their heads and a spot that looks like an eye on each side of their tails. The fish's real eye is in a dark, vertical stripe, which further confuses its enemies. The predator cannot tell which end of the fish is its head and which end is its tail. If a predator lunges toward the fish's tail in an attempt to capture it, the fish will probably have just enough time to swim straight ahead and escape.

Just as some harmless insects mimic those with painful stingers, a fish called *Calloplesiops altivelis* mimics the spotted moray eel, a well-armed reef fish. The moray eel has a dark body with small white and bluish dots, prominent eyes, and a mouth full of sharp teeth.

A copperhead butterfly fish has eyes on its head and spots that look like eyes on either side of its tail. This makes it hard for a predator to tell the head from the tail.

When *C. altivelis* feels frightened, it swims into the reef and leaves its rear end exposed. It then spreads out its tail fin and reveals an eyespot, so that it looks like the head of a moray eel. Most of the fish's enemies would rather not disturb these ferocious predators so they leave *C. altivelis* alone.

Some sea creatures rely on masking to discourage their enemies. One example is the decorator crab. All crabs are scavengers—they eat whatever they can find. Although some swim well, most walk or crawl with a peculiar sidelong gait that is easy for predators to recognize.

To make itself less noticeable, the decorator crab sticks tiny pieces of seaweed, sponge, and moss all over its body. These bits and pieces are held on by tiny hooked bristles on the crab's shell. When the crab is finished adorning itself, it looks like a rock.

Masking works well for the decorator crab, as it does for weevils and sloths. Just as masking is used by animals as different as insects and mammals, so are all the other types of camouflage.

What do many birds have in common with the copperband butterfly fish? How is an American bittern similar to a young tapir and a zebra? What does a snowshoe hare have in common with a bird called the rock ptarmigan? The answers to all these questions is the same—similar camouflage techniques.

A decorator crab is not so easy to recognize after it has attached other tiny sea creatures to its back.

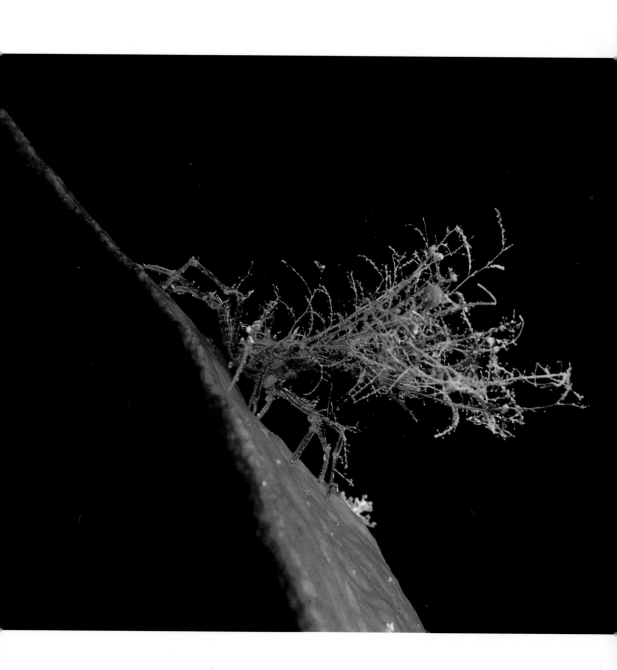

All over the world, animals rely on concealing coloration, countershading, masking, and mimicry every day of their lives. Some use camouflage to trick their enemies; others use it to sneak up on their prey. In the wild, camouflage is the best defense most animals have in nature's deadly game of hide-and-seek.

Glossary

amphibian—members of the class of vertebrates that includes frogs, toads, newts, and salamanders. Amphibians have moist, thin skin. Their eggs must be laid in water and hatch into larvae before becoming adults. Adults can live on land but must return to the water to reproduce.

camouflage—any form of disguise that helps an animal blend in with its surroundings so that it is not noticed.

carnivore—a meat-eating animal.

chromatophore—in cuttlefish, specialized pouches that contain yellow, orange-red, or black pigments.

concealing coloration—a type of camouflage in which the animal's body color matches its surroundings.

countershading—describes animals that are darker on top and lighter underneath to counteract natural shadowing.

disguise—describes animals that hide by resembling objects such as leaves, sticks, or rocks.

disruptive coloration—spots, stripes, or splotches of color that disrupt or break up the outline of an animal's body.

eft—a life stage of the newt during which the animal spends all of its time out of water.

habitat—the place where an animal or a plant spends its life.

lichen—a plantlike growth found on trees and rocks. It is a mixture of an alga and a fungus.

mammals—a group of animals that includes bears, cats, dogs, whales, and humans.

mantid—the name of a group of insects.

masking—when an animal hides or masks itself by adding objects to itself from the surrounding environment.

mimicry—when one animal copies the appearance, movement, sound, or smell of another animal.

nocturnal—active at night.

polymorphism—a type of camouflage in which the coloration of an animal varies depending on where it lives.

predator—an animal that hunts and eats other animals.

prey—an animal that is hunted and eaten by other animals.

reptile—a member of the class of animals that includes snakes, lizards, turtles, alligators, and crocodiles. Reptiles have dry, scaly skin and lay eggs.

For Further Reading

Duprez, Martine. *Animals in Disguise*. Watertown, MA: Charlesbridge, 1994.

Martin, James. *Hiding Out*. New York: Crown, 1993.

Pope, Joyce. *Mistaken Identity*. Austin, TX: Steck-Vaughn Library, 1992.

Powzyk, Joyce. *Animal Camouflage: A Closer Look*. New York: Bradbury Press, 1990.

Sowler, Sandie. *Amazing Animal Disguises*. New York: Alfred A Knopf, 1992.

Index

About the Author

Phyllis J. Perry has worked as an elementary schoolteacher and principal and has written two dozen books for teachers and young people. Her most recent books for Franklin Watts are *The Crocodilians: Reminders of the Age of Dinosaurs*, *The Snow Cats*, and *Armor to Venom: Animal Defenses*. She did her undergraduate work at the University of California, Berkeley and received her doctorate in Curriculum and Instruction from the University of Colorado. Dr. Perry lives with her husband, David, in Boulder, Colorado.